GEORGIA
AQUARIUM

BECKON BOOKS

TABLE OF CONTENTS

AT PEMBERTON PLACE
Georgia Aquarium sits on nine acres donated by the Coca-Cola Company. The land is named in honor of Dr. John Pemberton, who invented Coca-Cola in 1886.

THE WORLD'S LARGEST AQUARIUM

Georgia Aquarium has been certified by the *Guinness Book of World Records* as the largest
aquarium in the world, with more than 10 million gallons of water, 120,000 fish, and 604,000 square feet.

In early 2001, Bernie Marcus, co-founder of The Home Depot, asked renowned aquarium director Jeff Swanagan to meet with him. Swanagan was expecting a donation to one of his marine conservation projects. Instead, he received a job offer and the opportunity to help Marcus achieve his latest philanthropic goal: to build a world-class aquarium in landlocked Atlanta.

Months later, the two men stood with then-governor Roy Barnes in a vacant downtown Atlanta lot, surveying the spot for what would become the world's largest aquarium. Marcus gave Governor Barnes a blue bowl as a symbolic gift.

"Because we don't have any plans to show you yet, we're presenting you with this bowl as a loan," said Marcus. "In 2005, when we present you with a new aquarium, you can give us the bowl back."

Shortly thereafter, Marcus, Swanagan, and a group of community leaders embarked on a world tour of aquariums. The group traveled 109,000 miles and visited 56 aquariums in 13 countries, from the London Zoo's historic Fish House (the world's first public aquarium, built in 1853) to the spectacular Ring of Fire Aquarium

MARINE MASCOT
The Yellow-banded Sweetlips, above, one of the Aquarium's newest mascots, can be found in Tropical Diver.

CONSTRUCTION AND COLLABORATION
The world's largest aquarium was built in 28 months, left. During its development, Georgia Aquarium collaborated with groups throughout the state.

in Osaka, Japan (constructed in 1990). The founders returned with a clear vision: to design the Aquarium around animals first rather than architecture. In fact, they didn't even consider hiring an architectural firm until they knew exactly which animals and exhibits would be housed inside. Marcus selected the highly respected Atlanta firm Thompson, Ventulett, Stainback & Associates (TVS). "Don't get carried away designing an extravagant building," he instructed the architects.

The crew broke ground on May 23, 2003, and construction was completed in the fall of 2005. On November 23, 2005, the people of Atlanta got their first look at Bernie and Billi Marcus's $250 million gift, and their response was enthusiastic. The media declared the Aquarium to be the best of both worlds—a magnificent landmark for downtown Atlanta and a functional space dedicated to researching and conserving aquatic life from around the world.

Thanks to its generous benefactors, Georgia Aquarium opened with zero debt. In addition, thanks to the public's overwhelming response, the Aquarium was almost immediately self-supporting. During its first year, a record-setting 3.6 million people passed through its doors.

10 MILLION GALLONS AND STATE-OF-THE-ART SYSTEMS

Outside, Georgia Aquarium resembles a blue metal and glass ship breaking through silver waves; inside, the building immerses visitors in aquatic blue waters. The Aquarium is organized around a large atrium, which flows into six different aquatic galleries—Ocean Voyager, Tropical Diver, Georgia Explorer, River Scout, Dolphin Tales, and Cold Water Quest. Each gallery is designed to demonstrate the importance of water and aquatic ecosystems.

These galleries are supported by a sophisticated life support system. With 10 million gallons in more than 70 different exhibits, the Aquarium has the largest and most technologically advanced pump and filtration system in the world. There are more than 70 miles

BLUEGILL
Bluegills, above, are distinguished by a dark spot located near the base of their dorsal fins.

A $250 MILLION GIFT
In 2003, Bernie Marcus, right, donated $250 million to develop Georgia Aquarium. He hired Jeff Swanagan, far right, to be the first director.

»»·ᴄ·– SIZING UP –·ᴄ·««
Approximately 230 newly
constructed average-sized
American homes can fit inside
Georgia Aquarium.

INSIDE THE AQUARIUM
A nonprofit organization,
Georgia Aquarium
features more than 70
exhibits that are designed
to inspire, educate, and
entertain.

of pipe in the Aquarium, enough to encircle the city of Atlanta on the I-285 loop. In addition, each minute more than 500 pumps push 300,000 gallons of water through 187 sand filters, 76 towers, and 91 protein skimmers.

Highly automated, the Aquarium's life support system can make 150 million decisions per second through a network of 24 computers. Approximately 4,500 alarms automatically alert technicians to any abnormalities, while tight water temperature parameters in the exhibits are maintained to within tenths of a degree. The system also allows the Aquarium's team to manually control pumps, valves, and water flow from anywhere in the building.

This life support system is both high-tech and environmentally friendly. Each exhibit uses a closed loop system where the water is filtered, treated, and returned to the habitat. These advanced filtration and water reclamation techniques enable the Aquarium to recycle and recirculate more than 99.5 percent of its exhibit water volume each week, saving approximately 4.5 million gallons per year.

Touch the Heart to Teach the Mind

Swimming through the waters at Georgia Aquarium are many thousands of marine animals representing an estimated 500 species. These species range from the giant polka-dotted whale sharks of Taiwan (the largest fish in the world) to the delicate weedy sea dragons of Australia.

The animals are cared for at the Correll Center for Aquatic Animal Health—a 10,500-square-foot space with a surgical suite, intensive care units, commissary, scrub rooms, life support rooms, maintenance tech rooms, pathology records room, water quality lab, treatment and quarantine space, and diagnostic lab. Designed by world-class veterinary professionals, the Center partners with

PROTECTING WHALE SHARKS
In 2009, the Aquarium participated in research that led to the establishment of the Whale Shark Biosphere Reserve off the northern tip of the Yucatan.

SEA OTTER RESCUE
Through its 4R program, the Aquarium has adopted three southern sea otters, right, that could not be returned to their natural habitat.

the University of Georgia Veterinary Teaching Hospital to provide residents with hands-on research opportunities in wildlife medicine and veterinary pathology.

This focus on science, education, and conservation extends to visitors of all ages and is in keeping with one of the Aquarium's founding philosophies: "Touch the heart to teach the mind." Currently, Georgia Aquarium is the only aquarium in the United States to provide a "Learning Loop," a behind-the-scenes space that connects school-age students with some of the most interesting aquatic animals in the world.

In addition, the Aquarium has become a local conservation leader with its extensive internal recycling program and innovative employee incentives for adopting green practices. The Aquarium is also an international supporter of many conservation-related projects, such as the national Seafood Watch list, and it has had a long-term partnership with the Ocean Project, which provides guests with "a lasting, measurable awareness of the importance, value, and sensitivity of the ocean."

Through it all, Georgia Aquarium remains committed to being a world leader in the exhibition and conservation of aquatic wildlife. As the world's largest aquarium, it upholds a comprehensive mission: "To be an entertaining, engaging, and educational experience, inspiring research and an appreciation for the animal world."

ZEBRA SHARK
Female zebra sharks lay dark brown or purplish-black-colored eggs with a tough, leathery covering.

IN THE FIELD

With partners in the National Marine Fisheries Service, Georgia Aquarium helped pioneer health assessments for beluga whales in Alaska in 2008, using methods developed at the Aquarium. In 2011, Aquarium researchers focused on studies that have been done on whales in Bristol Bay— particularly the population in the Cook Inlets, which was recently listed as endangered. Based on blood samples and biopsies, the Aquarium studied the belugas' diet and tested for any exposure to pollution.

TURTLE TRACKING
Rehabilitated loggerhead sea turtles are fitted with satellite tracking devices to measure their progress upon release.

AN EXPANSIVE EXHIBIT
The filtration pipes in Ocean Voyager stretch for 61 miles, equal to the length of I-285 around Atlanta.

SIZING UP
It took 1.7 million pounds of sea salt to make the Ocean Voyager exhibit as salty as the ocean.

OCEAN VOYAGER
BUILT BY THE HOME DEPOT

Home to the world's largest fish—the only whale sharks exhibited outside of Asia—Ocean Voyager also boasts the world's largest aquarium pool, measuring 284 feet long, 126 feet wide, and 30 feet deep.

Schools of trevally jacks, spotted eagle rays, largetooth sawfish, enormous giant grouper, and several sharks inhabit the waters of Ocean Voyager, built by The Home Depot. The exhibit is specially designed to support whale sharks, the world's largest fish species. With its unique design and massive length, Ocean Voyager enables the animals in this habitat to navigate in different directions—a departure from traditional circular exhibits.

Ocean Voyager holds 6.3 million gallons of saltwater and is supported by walls that are up to four feet thick and more than 30 feet high. An intricate network of pipes supports the gallery's filtration systems, some of them as large as 54 inches in diameter. In addition, a dedicated computer system—continually monitored by highly trained staff—controls the pumps, valves, filters, and water flow, connecting 3,200 control points via 25 miles of wiring. With 4,574 square feet of viewing windows, a 100-foot-long underwater tunnel, and the second largest viewing window in the world, Ocean Voyager provides multiple locations to view some of the world's biggest and most unusual fish.

PORKFISH
Porkfish, above, are native to the western Atlantic. They are relatively unafraid of divers and can often be closely approached.

GROUP DYNAMICS
Blue tang surgeonfish, left, sometimes descend upon reef areas en masse to overwhelm other territorial fish.

WHALE SHARK
RHINCODON TYPUS

Ocean Voyager is the only aquarium outside of Asia and one of only a handful in the world to exhibit whale sharks. These sharks, however, are not whales. They get their name from their impressive size. As the largest fish in the ocean, whale sharks can reach lengths of 40 feet. They congregate for several months each year at about 20 locations throughout the world to feed on seasonal blooms of plankton. These sharks—which are harmless to humans—are listed as vulnerable on the IUCN Red List due to pressures from unregulated fisheries in China, India, and the Philippines.

Since they are not easy to find or study in oceans and very little is known about them, whale sharks are of great interest to scientists and conservation biologists. One of Georgia Aquarium's long-term goals is to understand their life span and natural history. To do this, the Aquarium has partnered with the five other facilities in Asia that care for them to study their growth, behavior, health, and genetics. Although scientists have not determined their life span, they do know these fish can live at least 12 years, since one whale shark has been on display in Japan for that length of time.

IN THE FIELD

Georgia Aquarium has been researching whale sharks in the field since 2005. One aspect of this research has been in the shallow coastal waters of the Yucatan Peninsula, near Cancun, Mexico. Aquarium scientists are focused on the number of whale sharks that gather there, why they come, and where they go once they leave the area. They are also studying an unprecedented gathering that took place in 2010, when hundreds of whale sharks congregated in a tiny patch of water to feed on fish eggs. Biologists hope to learn more about how the whale sharks were able to locate these relatively small amounts of food in the vastness of the ocean.

SPOTTED WOBBEGONG
This shark species, found in Ocean Voyager, is a master of camouflage.

WHALE SHARK
As the world's largest fish, adult whale sharks have skin that is four inches thick. The longest accurately recorded whale shark was 40 feet.

MANTA RAY
Manta rays are threatened by pollution, overfishing, climate change, predation, poachers, and accidental by-catch.

MANTA RAY
MANTA BIROSTRIS

Georgia Aquarium is one of just four aquariums in the world—and the only aquarium in the United States—to exhibit Manta rays. The largest rays in the ocean, Manta rays have broad heads and enormous mouths flanked by two wide, fleshy cephalic lobes. These animals are also called "devil rays," because when their cephalic lobes are rolled and projected forward, they take on the appearance of horns. Manta rays are ovoviviparous, meaning that the embryo develops within eggs retained in the mother's uterus. They give birth to one or two live young. Young Manta rays, called pups, are born with their wings folded around their bodies to allow for easier delivery. At birth, their wings measure about five feet across, growing to an average width of 13 feet. They can grow to reach 26 feet across and weigh up to 3,000 pounds.

Manta rays primarily feed on plankton, though they can also consume small- and moderate-sized fish. During feeding, they repeatedly somersault underwater, occasionally breaking the surface. The Aquarium's Manta rays often display this behavior. Manta rays do not have a stinging spine and are generally harmless to humans.

IN THE FIELD

Georgia Aquarium was the first aquarium in the U.S. to exhibit a Manta ray. Rescued from shark nets in South Africa in May 2007, the ray was initially very small at just eight feet across. The ray, named Nandi, was rehabilitated by uShaka Marine World, the largest marine park in Africa, and in just one year, she doubled her weight and outgrew her 580,000-gallon exhibit. In 2008, Georgia Aquarium and uShaka created an international partnership to bring the Manta ray on a 9,000-mile journey to her new home in Ocean Voyager, where the Aquarium hopes to raise worldwide awareness about the species. Georgia Aquarium has since acquired three additional Manta rays to join Nandi.

LEOPARD WHIPRAY
The tail of this stingray species—which occurs in the Indo-Pacific—can be three times the length of its body.

WHALE SHARKS

Along with a number of partners, Georgia Aquarium is researching whale sharks to work toward managing the sharks' worldwide population. The Aquarium's partners include the government of Taiwan and the Taipei Economic and Cultural Office in Atlanta. As a result of these efforts, Taiwan reduced its whale shark fishing quota from 60 in 2006, to 30 in 2007, to zero in 2008 and beyond. Georgia Aquarium is also sponsoring research on whale sharks with Mote Marine Laboratory in the Caribbean and National Taiwan Ocean University. Through these efforts, the Aquarium hopes to educate the public on aquatic conservation and encourage other countries to adopt sustainable seafood practices.

SEA CENSUS

The Aquarium is comparing genomics from its own whale sharks with free-ranging whale sharks in Mexico to see if there are differences in the populations.

ZEBRA SHARK
STEGOSTOMA FASCIATUM

Zebra sharks get their name from the vertical black-and-white stripes they have upon hatching. These stripes gradually break into leopardlike spots as they grow. Slow moving and nonaggressive, zebra sharks are found on coral reefs from the west Pacific to South Africa and the Red Sea. They feed on snails, shrimp, crabs, sea urchins, and small fish at night and rest on the ocean bottom during the day. These sharks can fit into small crevices and holes in the reef as they search for food.

Georgia Aquarium participates in the Species Survival Program (SSP) for zebra sharks, a program that strictly manages all zebra shark specimens to maximize their breeding potential. In 2006, the Aquarium's zebra sharks began producing large egg cases, and after a nine-month incubation period, the eggs began to hatch. Dozens of eggs have been hatched since, and some of those sharks are now on display at several other aquariums in the United States.

SIZING UP
The acrylic window at the Ocean Voyager exhibit is the largest window in North America, measuring 26 feet tall, 63 feet wide, and nearly two feet thick.

POTATO GROUPER
These fish, which can live 40 years, are territorial and aggressive toward divers. All are born female; some later become male.

SIGHT SEA-ING
Ocean Voyager has 185 tons of acrylic windows, right, to view the world's largest and most varied collection of sea animals.

SAND TIGER SHARK
Sand tiger sharks, far right, swim to the surface and swallow air in order to regulate their buoyancy. This enables them to remain motionless at any depth while seeking their prey.

ZEBRA SHARK
These bottom dwellers are fed via a device at the surface to provide them with vitamins and monitor their overall food consumption.

LIVING ART
The Tropical Diver gallery contains exhibits featuring sea nettles, jellies, seahorses, sea stars, anemones, giant clams, anthias, eels, and more!

TROPICAL DIVER
PRESENTED BY SOUTHWEST

Few places on the planet can match the complexity, diversity, and
beauty of life on a Pacific barrier reef.

Tropical Diver, presented by Southwest Airlines, contains one of the world's largest living coral reefs in an aquarium. With more than 180,000 gallons of water, this gallery is home to garden eels popping out of the sand, swirling masses of tiny glass sweepers, clownfish, seahorses, fairy basslets, palette surgeonfish, yellow-head jawfish, and more! Tropical Diver also features several exhibits of jellies from the Pacific.

The largest habitat in this exhibit, the Pacific Barrier Reef, is 49 feet long by 47 feet wide by 18 feet deep, and it contains more than 164,000 gallons of water. Only a small portion of this exhibit can be viewed from the general gallery. Special education groups and behind-the-scenes participants, however, can tour the gallery's upper level to see the habitat's complete wave zone, reef crest, lagoon, and living mangrove swamp. This upper level conveys how an entire reef ecosystem functions in nature and is only available on the Learning Loop.

LONGSNOUT SEAHORSE
Seahorses, above, pair-bond for life. The males give birth to up to 1,500 young at once.

UNDERWATER OASIS
Yellow-head jawfish, far left, and cherry barbs, left, are two of more than 100 fish species seen in Tropical Diver.

YELLOW-BANDED SWEETLIPS
PLECTORHINCHUS LINEATUS

Yellow-banded Sweetlips are native to the western Pacific from the Ryukyu Islands of Japan to Australia. They can often be found in the Pacific barrier reef, swimming among various species of anemones and corals. A member of the grunt family, Yellow-banded Sweetlips have a striking black-and-white pattern covering most of their body, yellow fins with black spots, an orange-red blotch on the base of their pectoral fins, and bright yellow lips. While their lips are large and fleshy, Yellow-banded Sweetlips actually have small mouths. They are typically nocturnal, bottom-feeding predators that feed on small fish and benthic crustaceans, such as shrimp. They can reach up to 28 inches long and produce low sounds using their pharyngeal teeth and swim bladder. While adults can be found either alone or in pairs, juveniles—which have much different coloration—are usually solitary.

YELLOW TANG
Found along the shallow reefs of the Pacific and Indian oceans, this species of surgeonfish removes algae from the shells of sea turtles.

RADIAL FIREFISH
Radial firefish, right, have venom-producing glands on their spines that can inflict a painful sting.

MOON JELLY
Unlike other species, moon jellies, far right, are true jellies. Their bodies are filled with a jellylike substance. They feed on phytoplankton.

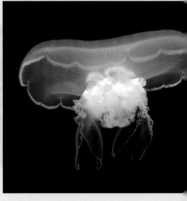

YELLOW-BANDED SWEETLIPS
This grunt species, which is typically active at night, develops spots on its fins when it reaches adulthood.

DID YOU KNOW?
Garden eels rarely leave their burrow and even reproduce by leaning over to intertwine with an adjacent mate.

GARDEN EEL
There are about 140 spotted (foreground) and splendid (background) garden eels at Georgia Aquarium. Some are almost a foot long.

SPOTTED GARDEN EEL AND SPLENDID GARDEN EEL
HETEROCONGER HASSI AND GORGASIA PRECLARA

Garden eels can be found in the tropical waters of the Indo-Pacific region from Africa to Japan to Australia. These fish usually live in dense colonies of up to several hundred individuals, extending just one-quarter of their 16- to 24-inch bodies above the surface of the sea floor. They use their pointed tails to burrow into the sand, secreting mucus to bind the sand grains along the walls of their burrows. From a distance, their thin bodies appear like a meadow of plants waving in the breeze, earning them the name "garden eels."

Two species of garden eels can be seen in Tropical Diver—spotted garden eels, which have spots, and splendid garden eels, which are striped. All garden eels in a colony will face into the current so they can pick plankton out of the passing water. At Georgia Aquarium, they usually face the left side of the exhibit, since the water and food flow from that direction. In their natural environment, garden eels quickly retreat below the surface of the sand at the approach of a predator. Among their predators are triggerfish, which dive into the eels' sandy burrows and dig them out when they try to hide.

A SPLENDID SPECIES
Garden eels, such as this splendid species, are born all black. They occur in dense groupings and appear like swaying plant stems.

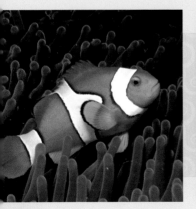

ANEMONE CLOWNFISH
These fish, far left, have a layer of mucus that makes them immune to the anemone's stinging tentacles.

BLUE TANG
Blue tang surgeonfish, left, change color as they mature, shifting from bright yellow to bright blue and finally to a deep purplish blue.

CORALS

orals are tiny animals that belong to the group *cnidaria*. They are not mobile but stay fixed in one location. They live in colonies consisting of many individuals called polyps. Stony corals secrete a hard calcium carbonate skeleton that provides a uniform structure for the colony and protects the polyps from predators. It is these hard skeletal structures that build coral reefs over time.

More than 3,000 species of fish and 500 species of corals live on Pacific coral reefs, making them one of the most diverse aquatic regions on the planet. Unfortunately, an estimated two-thirds of them are threatened as a result of pollution, overfishing, disease, and rising ocean temperatures. Most of the corals in Tropical Diver were cultivated from other aquariums through propagated coral "live rock." This was relatively easy to do, since many species of corals from the Pacific have been cultivated in home aquariums and public aquariums for more than 20 years.

IN THE FIELD

Georgia Aquarium and the Steinhart Aquarium at the California Academy of Sciences operate two of the largest living reef aquariums in the world. In order to best handle this technical challenge, these aquariums are collaborating on how to maintain thriving live reef exhibits on such a large scale. Georgia Aquarium also works with groups in the Solomon Islands and the Florida Keys to conserve coral reefs in the field. By conducting research at the Aquarium and in the field, the conservation and research team has made strides in saving these complex ecosystems.

PACIFIC SEA NETTLE
These jellies can have one- to three-foot domes and arms that extend as long as 12 feet from their undersides. They eat small fish, fish eggs, larvae, and even other jellies.

CORAL REEFS
While corals cover less than 1 percent of the ocean floor, they support about 25 percent of marine life.

A STATE OF FUN
The touchpools and interactive features in Georgia Explorer make the gallery a favorite area for children.

SUNTRUST GEORGIA EXPLORER

Highly interactive exhibits educate the Aquarium's guests about the future of Georgia's aquatic wildlife—from loggerhead sea turtles to endangered northern right whales to invasive lionfish.

Inhabiting the SunTrust Georgia Explorer gallery are animals native to the coast of Georgia, including a loggerhead sea turtle and an abundance of fish from Gray's Reef, a national marine sanctuary off the coast of Georgia. The gallery also educates guests about Georgia's state marine mammal, the northern right whale, one of the most endangered mammals on the planet. Most of the animals in the Georgia Explorer gallery are native, with one notable exception—the invasive lionfish. These fish are growing in numbers on the Georgia coast, and they have changed the makeup of the state's coastal ecosystem by consuming food that is normally available for native species.

The Georgia Explorer gallery also features interactive touch pools with rays, sharks, shrimp, sea stars, and other animals found along the coast of Georgia. One of the smaller species of hammerhead sharks, bonnethead sharks have heads that are more mallet-shaped than other hammerheads. Forbes sea stars, or common sea stars, are found from the Gulf of Maine to Texas. When feeding, they will wrap their five arms around their prey, gripping it with their suckered tube feet. In their natural habitat, they feed on mollusks.

ATLANTIC SPADEFISH
Atlantic spadefish, above, are often mistaken for angelfish. Adults often congregate in groups of up to 500.

PLEASE TOUCH
When interacting with the animals in the touchpools, left, guests are asked to use a delicate, two-finger touch.

LOGGERHEAD SEA TURTLE
CARETTA CARETTA

Typically found in the shallow, coastal waters of tropical and warm temperate seas, loggerhead sea turtles get their name from their massive heads. With their powerful jaws, they crush their prey, which includes hard-shelled, bottom-dwelling invertebrates such as conchs, clams, crabs, shrimp, oysters, and horseshoe crabs.

Loggerhead turtles spend the majority of their lives in the ocean. During the nesting season, females will crawl onto the beach to lay eggs in a nest scooped in the sand. The two-inch hatchlings emerge from the nest within 45 to 60 days and make their way to the ocean. Very little is known about where young turtles spend their early lives—they are rarely seen. As they grow older, they move closer to the coastline and hunt for food.

Loggerheads have been swimming in the oceans for millions of years, dating back to before the time of dinosaurs. Today, however, they are a threatened species. Loggerhead sea turtles have been overharvested for their shell and for food, and their nesting sites have been lost to development. Despite conservation efforts over the last 30 years, their numbers are in decline.

IN THE FIELD

Georgia Aquarium has partnered with the U.S. Fish and Wildlife Service and the Caretta Research Project to monitor several turtle refuges along the Georgia coast. The first loggerhead sea turtle to live at Georgia Aquarium was rescued from one of these sites. The turtle didn't emerge from his underground nest on Jekyll Island after hatching, so he was taken to the Tidelands Nature Center and then to the Aquarium. After almost two years at the Aquarium, the loggerhead was brought back to Jekyll Island, where he lived at the Georgia Sea Turtle Center and was trained to find food in the ocean.

BROWN SHRIMP
Adult and juvenile brown shrimp stay close to the ocean floor and burrow into the sand to escape predation.

LOGGERHEAD SEA TURTLE
Although sea turtles have no teeth, they have powerful jaws to crush their prey—primarily invertebrates. They rarely eat live fish.

RED LIONFISH
Lionfish have zebralike red-and-white stripes that vary depending on geographic location, habitat, and water depth.

RED LIONFISH
PTEROIS VOLITANS

Native to the Indo-Pacific region, red lionfish are widely distributed among coral and rock reefs and other tropical inshore habitats. They were introduced into the western Atlantic in the early 1990s and now are found in coastal waters from New York to South Florida, as well as Bermuda, the Bahamas, and parts of the Caribbean.

Although scientists do not yet fully understand the consequences of the lionfish invasion, they are concerned that these fish are causing significant changes to the ecosystems in these regions. One primary reason is that lionfish prey on and displace native fishes. Lionfish stalk their prey and corner it with their outstretched pectoral fins, seizing it in a lightening-quick lunge and then swallowing it whole. At night, they glide along the rocks and coral; by day, they hide under ledges, in caves, and in crevices.

Red lionfish can reach 15 inches long and weigh two-and-a-half pounds. They rely on their unusual colors and venomous spines to discourage predators. While courting, males are particularly aggressive and will attack interloping males by biting and pointing their venomous dorsal spines forward. Spawning takes place near the surface. The eggs hatch in about 36 hours, and the larvae drift in the plankton for weeks before settling to the bottom. This accounts for their wide native distribution and rapid spread in the western Atlantic.

GOLIATH GROUPER
This critically endangered fish, above, can reach eight feet long and weigh 1,000 pounds.

PENCIL URCHIN
Pencil urchins, left, have distinct thick, blunt spines that are used to wedge themselves into rocks during the day. These spines are helpful in areas with a strong current.

SEA TURTLES

Georgia Aquarium's Conservation Field Station (GACFS) in Florida has been actively involved in rehabilitating sea turtles along the Atlantic coast. In February 2010, 4,500 turtles were stranded in Florida as temperatures dropped below 50 degrees. With the help of GACFS, Georgia Aquarium was able to rescue several stranded loggerhead sea turtles. The animals were transported to the Aquarium's quarantine facility in Atlanta and the Georgia Sea Turtle Center, where staff treated and monitored the animals. All the turtles had lesions on their shells, heads, flippers, and necks; they were also severely underweight and malnourished. Veterinary staff and biologists cared for their wounds, drew blood, ran X-rays, provided antibiotic therapy, and monitored their conditions. Once veterinary staff members were confident the animals were healthy, they began introducing live food to their diets to ensure their predatory instincts would take over again when they returned to their natural environment. The turtles were released in the summer of 2010 and were fitted with scientific satellite tracking devices so their migration, behavior, and progress could be studied.

TRACKING THEIR PROGRESS
Since 2005, the Aquarium has rehabilitated and released eight loggerheads. Their behavior, migration, and progress are monitored through satellite tracking devices.

ROBUST REDHORSE SUCKER
MOXOSTOMA ROBUSTUM

ᨠᨠᨠ

Although most of the exhibits in the Georgia Explorer gallery display marine organisms, one exhibit highlights a rare freshwater fish: the robust redhorse sucker. These fish are found in Georgia, North Carolina, and South Carolina, where they live in silty to rocky river pools and in slow-flowing sections of small to medium rivers. Spawning occurs in coarse gravel habitats from late April to early July. Robust redhorse suckers can reach nearly 30 inches long and weigh 17 pounds. They are bronze on their backs and sides, and adults have faint stripes on their lower sides. Their lifespan is approximately 25 years.

This species was originally described in 1870, based on a six-pound specimen, but it was not rediscovered until 1991, when researchers collected a robust redhorse sucker from the Oconee River. Also called smallfin redhorse suckers, they are among the most threatened fish in North America and listed as endangered by the Georgia Department of Natural Resources. Today, a consortium of public and private companies and organizations are working with government agencies to preserve their habitat and augment natural populations with fish raised in aquaculture facilities.

BONNETHEAD SHARK
Relatively harmless to humans, these sharks, above, travel in groups of five to 15 same-sex individuals.

ROBUST AND RARE
The robust redhorse suckers at Georgia Aquarium, right, will eat right out of the aquarists' hands. They were raised on a fish farm.

ROBUST REDHORSE SUCKER

During spawning, robust redhorse suckers bury their eggs in gravel. The young fish remain there after hatching until they can swim.

FREE FLOW
Ninety-nine percent
of freshwater is frozen
in soil or locked in ice;
only 1 percent is free
flowing.

SOUTHERN COMPANY RIVER SCOUT

Healthy, free-flowing rivers are lifelines for our planet. They provide habitats for
more than 10,000 freshwater species and supply 60 percent of the drinking water in our homes.

The Southern Company River Scout gallery includes animals found in the rivers of Africa, South America, Asia, and Georgia. Throughout the gallery, a North American river flows overhead, providing visitors with a view of the Aquarium's bass, gar, and sturgeon fish. Albino alligators are also found in this gallery, along with electric eels, an emerald tree boa, archer fish, and Asian small-clawed otters.

Many of the animals featured in this gallery are disappearing in their natural environment due to pollution, loss of habitat, or overfishing. It is estimated that 20 percent of the 10,000-plus freshwater fish species in the world have become extinct or endangered in the last century, a number that is accelerating. Dams are a major factor in this habitat loss, since they change the ecology of rivers. More than 45,000 dams have been built around the world, leaving only 2 percent of the rivers in the United States and 60 percent in the world completely free flowing. In the United States, dams are being removed wherever possible to restore rivers to their natural state and to allow native fish to repopulate these waterways. Still, there is much more conservation work to be done.

RAISING THEIR YOUNG
Male red piranhas, above, make nests and guard the fertilized eggs until they hatch.

RUSHING WATER
River Scout, left, features a variety of fish, including elephantnose fish, electric eels, and mosquitofish.

ALBINO ALLIGATOR
ALLIGATOR MISSISSIPPIENSIS

Albino alligators are actually the same species as olive-and-black-colored American alligators. Their white coloring is due to a rare genetic mutation that affects the production of melanin, a skin pigment. There are two types of white alligators: albino alligators, which have red eyes (showing their underlying blood vessels), and leucistic alligators, which have blue eyes. Fewer than 50 albino American alligators live in the United States.

Albino alligators have different needs than normal American alligators. Sometimes called "ghosts of the swamp," they have an estimated survival rate of only 24 hours in the wild due to their sensitivity to direct UV radiation and their inability to blend in to their environment. At Georgia Aquarium, they are kept out of the sun, and their diet is supplemented with vitamin D_3.

The American alligator was first listed as an endangered species in 1967 due to loss of habitat and market hunting. A combined effort by the U.S. Fish and Wildlife Service, state wildlife agencies in southeastern states, and the creation of large, commercial alligator farms saved these animals from extinction. In 1987, the Fish and Wildlife Service announced the American alligator was fully recovered and removed it from the list.

> **⊷ SIZING UP ⊶**
> River Scout's 45,000-gallon overhead enclosure provides a perspective of rushing rivers that is unique among aquariums.

ELEPHANTNOSE FISH
These social fish, above, use their electric sensory receptors to hunt for worms and insects.

OVER THE RIVER
People everywhere depend on clean freshwater, and so do a myriad of fishes, other aquatic animals, and plants.

ALBINO ALLIGATOR
These American alligators have a genetic mutation that affects their production of melanin. In their natural environment, they have a slim chance of survival.

RED PIRANHA
Due to their razor sharp teeth, piranhas at the Aquarium are extracted for their medical exams in stiff plastic baskets—not nets.

PIRANHA
PYGOCENTRUS NATTERERI

Red piranhas are widely distributed in the basins of the Amazon, Paraguay-Parana, and Essequibo rivers. These South American freshwater fish prefer areas with dense vegetation, such as creeks and interconnected ponds. Piranhas are efficient eaters. They don't chew their food, and they constantly rotate places during meals to ensure that each member of their group gets the same amount of food. In their natural environment, their feeding times vary by age and size: Medium-sized individuals are most active at dawn, in the late afternoon, and at night, while smaller fishes feed throughout the day. Not all their feeding is violent. They are also known to scavenge for insects, snails, worms, and plants.

Adults can reach 8 to 12 inches long and have powerful jaws that contain sharp, triangular, interlocking teeth. Their predators include crocodiles, some birds, large catfish, and large mammals, such as jaguars. Humans also eat them or use them as bait for large catfish.

DID YOU KNOW?
Red piranhas rarely attack humans unless blood is in the water. At Georgia Aquarium, staff members dive in the piranha exhibit every week. The animals are not aggressive toward the divers and usually keep their distance.

CICHLID
Pronounced "sick-lids," these fish, above left, comprise one of the largest families in the world, with 1,350 species identified.

ELECTRIC EEL
Electric eels, left, have a special organ that produces a charge used for protection, communication, and navigation.

ASIAN SMALL-CLAWED OTTER
AONYX CINEREA

gile and active, Asian small-clawed otters dwell in the freshwater wetlands and mangrove swamps of southeast Asia, southern India, southern China, and the Philippines. Of the 13 species of otters in the world, they are the smallest, reaching only 18 to 24 inches long and weighing just six to 12 pounds. These carnivores eat mollusks, fish, and frogs, as well as crabs and other crustaceans.

Asian small-clawed otters spend more time on land than any other otter. They have narrow, nonwebbed feet, making them more agile on land but less efficient when swimming underwater. Their tail helps propel them through the water, and their whiskers can detect changes in water current and pressure. While swimming, they can seal their nostrils and ear canals and reduce their heart rate and oxygen consumption.

Small-clawed otters produce a strong, musty odor from the scent glands at the base of their tails. They use their scent to mark their territory and communicate their identity, sex, and sexual state. Small-clawed otters are threatened by habitat loss, hunting, and pollution. They are listed as vulnerable on the IUCN Red List.

ARCHERFISH
Archerfish, above, are known for shooting down insects with a stream of water from their mouth.

FROM MAMMALS TO FISH
Asian small-clawed otters, right, catch fish with their paws rather than their mouths. Discus, far right, are named for their circular, compressed bodies.

MAKING A MARK
The otters at the
Aquarium scent-mark
the rocks and trees in
their exhibit whenever
they are awake and
active.

THE WORLD'S A STAGE
The AT&T Dolphin Tales show was developed by award-winning producers and includes five original songs recorded by a 61-piece orchestra.

⊰⊱⊰⊱ **SIZING UP** ⊰⊱⊰⊱
Georgia Aquarium's exhibits hold the equivalent water volume of more than 100 million cans of soda!

AT&T DOLPHIN TALES

Bottlenose dolphins serve as a keystone species for monitoring
highly populated coastal areas and the overall health of our oceans.

Dolphins have been around for thousands of years and have been mentioned in the writings of Aristotle and other ancient thinkers. There are more than 30 species of dolphins— approximately 26 marine species and four river dolphins. AT&T Dolphin Tales was designed to both educate and entertain visitors about the most common species in zoos and aquariums: the bottlenose dolphin.

This gallery includes a lobby with a 25-foot underwater viewing window. Here, visitors can see the dolphins up close and observe their naturally playful behaviors, watching as they jump into the air and chase one another through the water. Staff and dolphin trainers are often nearby to share interesting facts about the animals. (For example, a bottlenose dolphin can hold its breath for eight to 10 minutes!) At the end of the entrance lobby is the Dolphin Tales Theater, which features a 30-minute performance that highlights the strong emotional bond between dolphins and humans.

BOTTLENOSE DOLPHIN
These dolphins, above, are often seen traveling with pilot whales, humpback whales, and other dolphin species.

A SOARING SPECTACULAR
More than two years in the making, the AT&T Dolphin Tales show, left, highlights the beauty and agility of bottlenose dolphins.

BOTTLENOSE DOLPHIN
TURSIOPS TRUNCATES

N amed for their distinctive elongated snouts, bottlenose dolphins are the best-known members of the dolphin family. These curious and sociable animals are found in most tropical to temperate oceans. Bottlenose dolphins range in color from bluish-grey to nearly black and have distinctive white undersides. Adults can measure between six and 12 ½ feet and weigh 330 to 1,400 pounds. Residents of warmer, shallower coastal waters tend to be smaller than those that migrate in cooler, deeper parts of the ocean.

Males compete for females during the breeding season, with the largest males at the top of the hierarchy. Courting involves rubbing, jaw clapping, stroking, and nuzzling between males and females. The gestation period for bottlenose dolphins is approximately 12 months. Newborn calves measure three to four feet, and birth weights can range from 30 to 50 pounds. Mothers and their calves remain closely associated for about four to five years.

Bottlenose dolphins live in groups—or pods—of up to 20 individuals, but large herds of several hundred can

IN THE FIELD

Bottlenose dolphins in Florida's Indian River Lagoon provide excellent cues to the area's environmental health. Since these dolphins are permanent residents of the lagoon and are at the top of the food chain, they can give insight into the health of the entire ecosystem. Dr. Gregory Bossart has spent years studying dolphins and the threats they face from pollution and emerging infectious diseases. He is working with other researchers in partnership with Florida Atlantic University and the federal government to understand the health of these animals. Georgia Aquarium's Conservation Field Station in St. Augustine plays a vital role in this research. This research was conducted under NOAA, National Marine Fisheries Service Permit Nos. 998-1678-01 and 14352-01, issued to Dr. Bossart.

SOCIAL STRUCTURE
Bottlenose dolphins live in groups of resident communities called pods.

BRAIN MATTERS
Bottlenose dolphins have ten times the brain area dedicated to acoustical imaging than humans. Humans have 90 percent more visual imaging.

be found offshore. They are often seen traveling with pilot whales, humpback whales, and other dolphin species. Bottlenose dolphins are known for their wide range of vocalizations, including whistles, grunts, squeaks, and moans. These vocalizations enable them to communicate with other dolphins in order to hunt efficiently, raise their young, and guard against predators. Some scientists believe bottlenose dolphins have a complex language and that individuals have their own distinctive whistle that provides information to other dolphins about their identity, location, and condition.

While they have very little sense of smell, bottlenose dolphins have excellent hearing: Their brains have 10 times more space dedicated to acoustical imaging than a human's. Like bats, dolphins also have a sophisticated system of echolocation. Nasal sacs in their foreheads enable them to produce high-frequency clicking sounds—up to 1,000 clicks per second! The bounce back from these signals is then received by fat-filled cavities in their lower jawbones and transmitted to their brains. With echolocation, dolphins can locate prey, identify predators, and navigate in dark or murky water. In addition, echolocation enables them to determine the size, shape, speed, distance, direction, and even the internal structure of objects in the water.

Dolphins usually hunt together, eating an estimated 15 to 30 pounds of food daily in their natural environment. Although they have 18 to 26 pairs of teeth in each jaw, they swallow their food whole. Their diet includes a wide variety of animals, particularly bottom-dwelling fish, shrimp, and squid. While they are not endangered, dolphins are vulnerable to getting caught in commercial fishing nets. Dolphins were once hunted for meat and oil, but they have been protected in the United States under the Marine Mammal Protection Act since 1972.

DID YOU KNOW?
Young dolphins develop a signature whistle and maintain it throughout their life.

CENTER STAGE
Generations of Americans have learned more about bottlenose dolphins through entertaining and educational shows in aquariums and oceanariums. They are currently the most common dolphin species in human care.

HOLDING THEIR BREATH
Bottlenose dolphins can remain submerged for an average of eight to 10 minutes.

A DIVERSE EXPERIENCE
Cold Water Quest features beluga whales, weedy sea dragons, southern sea otters, harbor seals, and African penguins.

✦ SIZING UP ✦
With 4,300 tons of cooling capacity, Georgia Aquarium's heating and air conditioning system could cool more than 1,400 average-sized homes—an important feature for the inhabitants of Cold Water Quest.

GEORGIA-PACIFIC COLD WATER QUEST

In terms of nutrients, cold waters are four times more productive per acre
than other parts of the world's oceans, enabling them to support a wide variety of aquatic life.

Georgia-Pacific Cold Water Quest is home to animals that inhabit the cool ocean waters from around the world. Some of the Aquarium's most popular animals are in this gallery, including beluga whales, harbor seals, southern sea otters, and African penguins. In addition, smaller animals—such as Australian weedy sea dragons, Garibaldi damselfish, and Japanese spider crabs—lurk beneath the waves, living among the rocky ledges and kelp forests.

Cold Water Quest has some memorable features. For example, the sea otter exhibit includes deck space for trainer-animal interactions and an underwater "clam cannon," one of many enrichment devices. The African penguin exhibit contains more than 25 nesting areas integrated into natural-looking rockwork and a lighting system that mimics the cycle of light from dawn to dusk.

Unfortunately, many cold water marine mammals and sea birds are in peril, threatened by overfishing and the effects of unusually warm sea temperatures. For some species, only the efforts of conservation organizations, committed individuals, and government officials will save them from extinction.

GARIBALDI DAMSELFISH
The largest of the damselfish, the Garibaldi, above, is the inspiration for the character of Deepo, the original mascot for Georgia Aquarium.

MAKING A SPLASH
Southern sea otters, left, can spend their entire lives in the ocean—though they are able to move on land.

BELUGA WHALE
DELPHINAPTERUS LEUCAS

The word "beluga" is derived from the Russian word for "white." Beluga whales live in the north polar regions of Alaska, Canada, Greenland, and northern Europe. There are an estimated 200,000 beluga whales. While they are not an endangered species, some local populations have been in steady decline for many years. Hunting was once a major threat, but today pollution poses a more serious concern.

The most vocal of the toothed whales, beluga whales can make at least 11 different vocalizations, including high-pitched whistles, squeals, clucks, mews, chirps, and bell-like tones. Arctic fishermen say they can hear beluga whales from miles away and feel the vibration of their sounds coming through the hulls of their fishing boats. This behavior has earned them the nickname "sea canary."

Beluga whales are extremely social, living in groups called pods. They hunt together to find fish,

IN THE FIELD

Georgia Aquarium has partnered with the Southern African Foundation for the Conservation of Coastal Birds (SANCCOB), a hands-on seabird hospital in Cape Town, to help preserve free-ranging African penguins like those in Cold Water Quest. These penguins live on a handful of offshore islands between Namibia and Port Elizabeth, South Africa. Recently reclassified from vulnerable to endangered, they consist of fewer than 26,000 breeding pairs—a number that has dropped by more than 80 percent in the last 50 years. African penguins are under severe threat from climate change, over-fishing of their major food sources, and predation. They are also particularly vulnerable to oil refineries and spills. Together, SANCCOB and Georgia Aquarium are rescuing, rehabilitating, and releasing wild African penguins back to their colonies to bolster the populations in the natural environment.

GIANT PACIFIC OCTOPUS
Octopuses have three hearts that pump blue blood. Each of their arms can have up to 1,800 suckers.

BELUGA WHALE
Beluga whales are the only whales with flexible necks, which help them capture their prey.

DID YOU KNOW?
Beluga whales undergo a seasonal molt by rubbing up against hard objects like rocks to shed their outer layer of skin.

DEEP SEA DIVERS
Belugas prey on approximately 100 kinds of animals, suctioning their food from the bottom of the sea with their lips.

octopus, squid, crabs, and other marine animals. While they have excellent eyesight, they often use echolocation to find their food, projecting broadband pulses with high frequencies from their foreheads and listening for returning echoes.

Beluga whales are mammals: They have lungs and must surface periodically to breathe through the blowholes on the tops of their heads. The blowhole is covered by a muscular flap that provides a watertight seal when they dive beneath the surface. While beluga whales typically breathe two to three times per minute, they can hold their breath for up to 25 minutes. With their flexible necks, belugas can swim upside down while they bend toward the surface to find a pocket of air or a breathing hole. Beluga whales also have the ability to purse their lips, which allows them to suck prey from the bottom of the ocean.

Belugas can reach 10 to 15 feet in length, weigh as much as 3,500 pounds, and live between 10 and 20 years in both an aquarium and an ocean setting. Breeding occurs from March through May, with gestation lasting about 14 to 16 months. Females give birth to a single calf in the late spring or early summer. At birth, calves are between three to five feet long and can weigh 100 to 200 pounds.

There are more than 30 beluga whales in human care in the United States. Since there are so few beluga whales in aquarium settings, breeding is carefully controlled to ensure their long-term viability in zoos and aquariums. As one of six U.S. aquariums and zoos that care for beluga whales, Georgia Aquarium is dedicated to preserving the species through conservation and research programs. The Aquarium participates in cooperative management and breeding programs as well as educational initiatives that inspire others to care about these fascinating animals.

AFRICAN PENGUIN
African penguins, above, are endangered. Their population has dropped 80 percent in the past 50 years.

WHITE WHALES
Beluga whales, left, can swim backward to maneuver through arctic ice. They have acute vision both in and out of the water.

SOUTHERN SEA OTTER
ENHYDRA LUTRIS

Sea otters are among the largest members of the weasel family and the only ones that live almost entirely in the water. They can eat, groom, rest, and nurse their young while they are floating. Sea otters live together in groups called "rafts," often holding paws while sleeping to keep from drifting away. Individuals usually swim on their backs but have been known to swim on their stomachs while traveling.

Their diet includes a variety of seafood: worms, sea urchins, abalone, mussels, scallops, octopus, squid, crabs, and fish. Sea otters may dive as deep as 200 feet for up to four minutes in search of food. Using rocks to crack open their prey, they consume 20 to 25 percent of their body weight each day to maintain their high metabolism.

As a keystone species, sea otters play an integral role in the ecosystem and health of our oceans. They help maintain the balance among thousands of kelp forest inhabitants. They also act as an indicator species, providing insight into the health of a marine ecosystem.

Sea otters were hunted nearly to extinction for their prized pelts. They are listed as endangered on the IUCN Red List and are protected by the Endangered Species Act and the Marine Mammal Protection Act. Today, about 2,200 survive in their natural environment.

GIANT JAPANESE SPIDER CRAB
Japanese spider crabs, above, live at depths of up to 1,000 feet and can live as long as 100 years.

LOBSTER
These 10-legged crustaceans, right, have poor eyesight but a highly developed sense of taste and smell.

DID YOU KNOW?
Sea otters sleep on their backs while floating on the surface of the water and wrap themselves with kelp to keep from drifting away.

SOUTHERN SEA OTTER
Southern sea otters stash their prey in pockets of skin under their front limbs. This leaves their paws free while they hunt.

SOUTHERN SEA OTTERS

Georgia Aquarium is committed to the rescue and rehabilitation of stranded southern sea otter pups off the coasts of California and Alaska. Only 25 percent of pups survive the first year, and when pups are separated from their mothers, the odds of survival drop dramatically. By working with groups like the Alaska SeaLife Center and Monterey Bay Aquarium's Sea Otter Research and Conservation (SORAC) program, Georgia Aquarium is aiding in the rescue of sea otter pups. One of the sea otters in the Cold Water Quest exhibit was the first to be successfully raised to adulthood in human care. To mimic their natural feeding behaviors, the otters are fed six to eight times every day; they are also offered a variety of enrichment devices to keep them mentally active and stimulated.

WELCOME HOME
Georgia Aquarium is home to four rescued sea otters. They were found stranded along the central California coast.

DID YOU KNOW?
Harbor seals have blunt snouts, and their V-shaped nostrils stay closed while swimming. This adaptation enables them to stay submerged for up to 40 minutes and dive to 1,450 feet.

HARBOR SEAL
When hauling out, harbor seals are on constant alert. They will quickly rush back into the water when alarmed.

HARBOR SEAL
PHOCA VITULINA

ᨒᨒ

Harbor seals live in the temperate, subarctic, and arctic waters of the North Atlantic and North Pacific Oceans. They are the most widely distributed of all pinnipeds (walrus, seals, and sea lions), inhabiting the shallow, near-shore waters of bays and estuaries where sandbars and beaches are uncovered at low tide so they can haul out and rest. Typically, harbor seals divide their time equally between swimming in the sea and resting on land.

Harbor seals are born with oil glands in their skin that help waterproof their fur, which ranges from silver-gray to black or dark brown. They are typically the least vocal seals, primarily communicating with each other underwater.

Harbor seals have prominent eyes that are adapted to seeing shades of black and white. Compared to humans, they have superior vision underwater. However, their whiskers are better for finding food. They detect their prey in low-light conditions by touch, vibration, or water movement.

Mothers generally give birth to one pup, which weighs around 20 to 24 pounds and is approximately two feet long. Within an hour of birth, the pup can swim and will soon ride on its mother's back. A pup will more than double its weight in the first month.

SOLITARY SEALS
Unlike most pinnipeds, harbor seals, above, are solitary and rarely interact except to mate.

WOLF EEL AND ANEMONE
Wolf eels, far left, are docile, attacking only when provoked. Though they look like plants, anemones, left, are animals that depend on tides to bring them food.

WEEDY SEA DRAGON
PHYLLOPTERYX TAENIOLATUS

Weedy sea dragons are found in shallow coastal waters among the kelp forests and reefs of southern and western Australia, including the southern tip of Tasmania. These slow, graceful swimmers use their olive green, yellow, purple, and blue coloration and their leafy appendages to camouflage themselves among the seaweed and sea grasses. They can be difficult to find as they slowly sway back and forth with the current.

Sea dragons are related to sea horses and pipe fish with the same long snouts and armorlike plates covering most of their bodies. Their snouts function like a straw, sucking up small crustaceans and other prey. The color, size, and shape of their leafy appendages depend on many factors, including food supply.

Georgia Aquarium is one of the only facilities in the world to have hatched and raised weedy sea dragons. Because sea dragons are so unusual, they are in great demand. Illegal collectors have threatened their populations, and the Australian government has put measures into place to protect them. The species is listed as endangered on the IUCN Red List.

SIZING UP
The African penguin exhibit in Cold Water Quest features a state-of-the-art lighting system that utilizes 44 independently controlled, high-output florescent and LED lights. The system follows the sunrise-to-moonset variations for the southern hemisphere, recreating the natural light cycle in the penguins' native South African habitat.

BIRDS OF A FEATHER
African penguins have up to 300 feathers per square inch. These feathers have waterproofing qualities that keep their skin dry.

DID YOU KNOW?
Male weedy sea dragons are the ones that give birth. While the females lay up to 300 eggs on the soft bottom of the males' abdomens, the males carry the eggs until they eventually hatch.

WEEDY SEA DRAGON
Juvenile weedy sea dragons have voracious appetites, eating more than 200 tiny mysid shrimp each day.

**MARINELAND
DOLPHIN ADVENTURE**
Marineland was
founded in 1938 to
provide a "window to
the ocean."

MARINELAND DOLPHIN ADVENTURE

In 2011, Florida's historic oceanarium, Marineland,
joined the Georgia Aquarium family to create Marineland's Dolphin Adventure.

ᙙᙚ

Located just south of St. Augustine, Florida, Marineland Dolphin Adventure opened in 1938 as Marine Studios. As the world's first oceanarium, the facility provided a place for Hollywood to capture underwater footage for TV shows and movies ranging from the *Tarzan* series to *The Creature from the Black Lagoon*. Later, the TV special *Benji at Marineland* featured the famous dog SCUBA diving, another first.

In the 1950s, the studio began to change focus, producing daily shows in which the dolphins performed to music. The formula was a success and became the model for many other aquariums, oceanariums, and marine parks worldwide. Yet while the facility was training animals to entertain the public, behind the scenes its researchers were pioneering studies in marine science, animal behavior, and water chemistry. The scientists at Marineland were the first to successfully breed Atlantic bottlenose dolphins, as well as to research and record dolphin echolocation, social behavior, and communication.

In 2006, the facility changed its name to Marineland's Dolphin Conservation Center. The old steel pools used primarily for the public's entertainment were replaced with new pools that better suited the animals' behavioral needs. Two years later, Georgia Aquarium established its

AN INTERACTIVE EXPERIENCE
Marineland Dolphin Adventure provides a variety of ways for guests to meet Atlantic bottlenose dolphins, from up-close experiences poolside to immersive, in-water interactions.

first conservation field station at the site, and in 2011, the Aquarium acquired the oceanarium and renamed it Marineland Dolphin Adventure.

While still serving as one of northeast Florida's premier attractions, the new Marineland Dolphin Adventure focuses on creating meaningful interactions with the dolphins and educating the community on ways to help preserve and protect these incredible animals in their natural habitat.

Today, programs at Marineland Dolphin Adventure enable visitors to enter the watery world of bottlenose dolphins and interact with these intelligent and social animals. In addition, there are plenty of opportunities to observe the dolphins as they interact with program participants, play with enrichment items, and work with researchers.

GEORGIA AQUARIUM'S CONSERVATION FIELD STATION

In 2008, Georgia Aquarium established the Georgia Aquarium Conservation Field Station (GACFS) adjacent to Marineland Dolphin Adventure. This state-of-the-art facility is fully equipped to respond to marine animal medical emergencies and to perform field studies.

As a member of the National Marine Fisheries Service's Marine Mammal Stranding Network, GACFS responds directly to any marine mammal strandings in Flagler County, Florida. In addition, the station assists participants in the neighboring southeast region in stranding and rescue operations. GACFS is also involved with the outreach and photo identification of bottlenose dolphins and other marine mammals.

In addition, GACFS actively participates in conservation outreach involving Antillean manatees and Amazonian manatees in Mexico, Belize, South America, Dubai, and at sites in Colombia and Brazil. Through these outreach programs, GACFS staff members provide experienced veterinary care to aquatic mammal species while training local caregivers to provide future care. Ongoing research programs like these seek to better understand why marine mammals strand and to assess, on a larger scale, emerging species diseases and the health of oceans worldwide.

WATER WORLD
The dolphins at Marineland Dolphin Adventure live in a 1.3-million-gallon facility designed for their behavioral needs. Several large pools also facilitate animal care and training.

SIZING UP

Marineland Dolphin Adventure is home to the oldest dolphin in human care, Nellie, born at Marineland on February 27, 1953. In contrast, the average lifespan of a bottlenose dolphin is 25 to 30 years.

CREATURE CONNECTION

By connecting people with dolphins, Marineland Dolphin Adventure aims to provide a better understanding of the species and inspire conservation efforts.

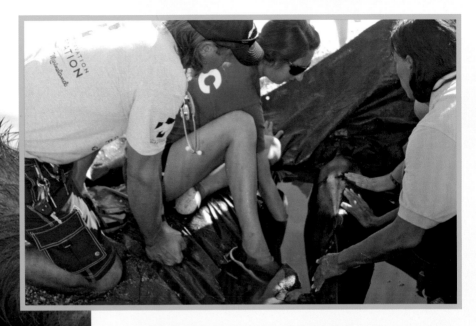

CONSERVATION FIELD STATION

Georgia Aquarium's Conservation Field Station (GACFS) has partnered with personnel and investigators from the Harbor Branch Oceanographic Institute who are conducting a population assessment study in the northeast Florida intracoastal waterways and Atlantic seaboard. Using photos of dorsal fins, they are collecting baseline information to help identify individual Atlantic bottlenose dolphins. A catalog of information will be created for animals in the area and will be compared with other known animals from other research areas on the East Coast. Information such as habitat usage, home range, and estimated population size will help lay the foundation for continuing research in the area.

CONSERVATION EDUCATION
The Aquarium supports dolphin conservation efforts at its field station in Florida through research, rescue, and rehabilitation.

GENTLE GIANTS
Georgia Aquarium has researched whale sharks with partners including Emory University, Georgia Tech, and the government of Mexico.

A SEA OF SUPPORT

Through its 4R Program, the Aquarium works to support and
fund efforts in the areas of Rescue, Rehabilitation, Research, and Responsibility.

Georgia Aquarium is making a difference in aquatic science around the world. Through its 4R Program, the Aquarium has helped to conserve many species, including beluga whales, loggerhead sea turtles, living coral reefs, and sea otters. The 4R Program has also played an active role behind the scenes by funding work through the Correll Center for Aquatic Animal Health; integrating a postdoctoral veterinary residency program in clinical medicine and pathology; establishing a state-of-the-art commissary; supporting Aquarium biologists; and conducting international research.

This 4R program is made possible through the generous support of Aquarium members. There are several levels of participation:

In the elite Oceans Society, members support aquatic animal research and conservation. These members receive numerous benefits such as annual passes, Behind-the-Scenes tours, special quarterly events, and the opportunity to swim with gentle giants— the whale sharks!

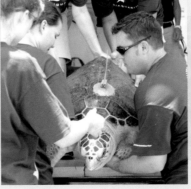

OTTER AID
One of the Aquarium's key projects is the rescue and rehabilitation of stranded southern sea otter pups, above.

RESEARCH AND REHABILITATION
Georgia Aquarium researches rare and endangered animals, far left, and rehabilitates stranded sea turtles, left.

For those who feel a special connection to whale sharks, the Aquarium offers its ADOPT program in which participants can adopt a plush whale shark. This purchase helps provide care for the world's largest fish at Georgia Aquarium and around the globe.

Atlanta's up-and-coming young professionals (ages 25 to 40) can support the Aquarium through its Next Wave Society. The society gives professionals the option to network with other emerging leaders, increasing their philanthropic support and conservation awareness. Membership provides environmentally focused volunteer opportunities, creative fundraising projects, and exclusive invitations to Next Wave Society events.

Finally, the Legacy Society works with individuals who establish planned gifts by including the Georgia Aquarium as a beneficiary in their estates. Legacy Society members receive individualized estate-planning assistance for bequests, trusts, IRAs, income-for-life plans,

and other charitable gifts to the Georgia Aquarium; invitations to exclusive donor events, special exhibit previews, Breakfast-with-the-Biologist events, and private Behind-the-Scenes tours; personal assistance with VIP ticketing and parking; and recognition in the annual report.

From the Oceans, Legacy, and Next Wave societies to the ADOPT program, each level of support will help ensure that Georgia Aquarium remains at the forefront of aquarium research and science for years to come.

IN THE FIELD

Georgia Aquarium is partnering with scientists from Woods Hole Oceanographic Institution to monitor right whale populations in the state. The North Atlantic right whale—the official mammal of the state of Georgia—is classified as endangered on the IUCN Red List. Right whales were once thought to be extinct; today, an estimated 300 remain. These whales breed every year in the warm waters of the South Atlantic Bight, which includes the Georgia coast. The greatest threats to the species are collisions with ships and entanglements in fishing gear. Efforts are being made by the government, conservation groups, and private citizens to minimize both hazards.

FORBES SEA STAR
Sea stars, above, can regenerate an arm, called a ray, if they lose one.

BELUGA EXAM
Staff members at the Aquarium's Correll Center for Aquatic Animal Health swab a beluga whale during a routine exam, right.

MAJESTIC MARINE LIFE
Aquarium laboratory staff test the water quality twice a day to ensure it is safe for all aquatic residents.

BECKON BOOKS

Georgia Aquarium in Atlanta, Georgia, is the world's largest aquarium, with more than 10 million gallons of water and one of the largest collections of aquatic animals anywhere. Centrally located in downtown Atlanta near world-famous Centennial Olympic Park, Georgia Aquarium welcomes well over two million guests each year, and more than 100,000 students who participate in focused learning programs. The mission of Georgia Aquarium is to be an entertaining, engaging, and educational experience, inspiring stewardship in conservation, research, and the appreciation for the animal world.

Georgia Aquarium is an accredited member of the Assocation of Zoos and Aquariums (AZA) and the Alliance of Marine Mammal Parks and Aquariums (AMMPA).

Georgia Aquarium is a 501(c)3 organization, and we rely on community support to fund our special programs, including education and veterinary services.

Special thanks to all of the Georgia Aquarium team members and volunteers, whose dedication and commitment to aquatic animals inspired and contributed to the publication of this book.

Unless otherwise noted, all photos are property of Georgia Aquarium.
For additional information, visit www.georgiaaquarium.org.

Georgia Aquarium
225 Baker St. NW, Atlanta, GA 30313
404-581-4000
facebook.com/georgiaaquarium

Georgia Aquarium was developed by Beckon Books in cooperation with Georgia Aquarium and Event Network. Beckon develops and publishes custom books for leading cultural attractions, corporations, and nonprofit organizations. Beckon Books is an imprint of Southwestern Publishing Group, Inc., 2451 Atrium Way, Nashville, TN 37214. Southwestern Publishing Group, Inc., is a wholly owned subsidiary of Southwestern, Inc., Nashville, Tennessee.

Christopher G. Capen, *President*, Beckon Books
Monika Stout, *Design/Production*
Betsy Holt, *Writer/Editor*
www.beckonbooks.com
877-311-0155

Event Network is the retail partner of Georgia Aquarium and is proud to benefit and support the Aquarium's mission of conservation, education, science, and recreation.
www.eventnetwork.com

ISBN: 978-1-935442-15-8
Printed in Canada
10 9 8 7 6 5 4 3 2 1